Entomologist Danny

written and photographed
by
Mia Coulton

Look at the bee.

1, 2, 3, 4, 5, 6...

The bee has six legs.

The bee is an insect.

Look at the ant.

1, 2, 3, 4, 5, 6...

The ant has six legs.

The ant is an insect.

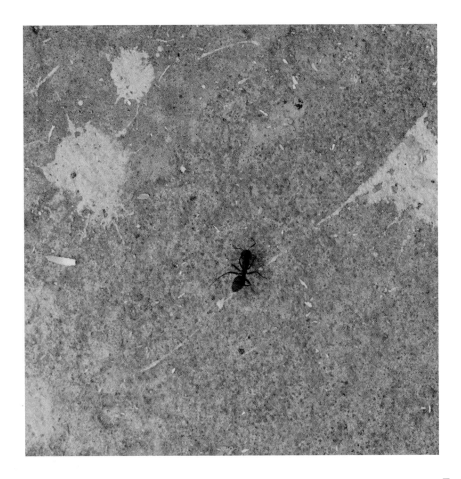

Oh my!

It's a spider.

1, 2, 3, 4, 5, 6, 7, 8!

The spider has eight legs.

It is not an insect!

The spider

is not an insect.

I do not look at spiders.

1, 2, 3, 4, 5, 6...

The butterfly has six legs.

The butterfly is an insect.

I look at insects.

I am a scientist.

I am an entomologist.